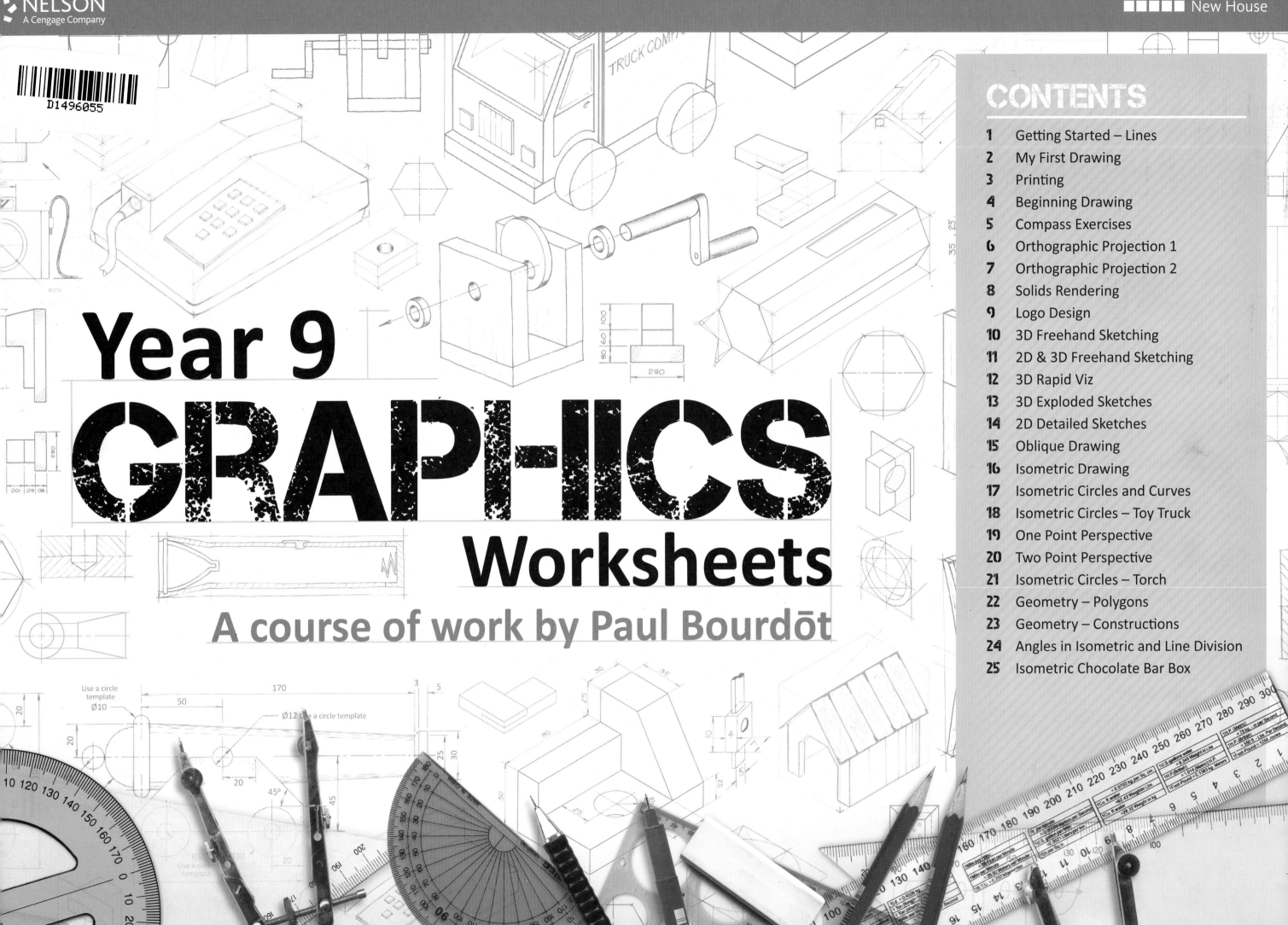

NAME:

Getting Started – Lines
Textbook page 11

My First Drawing

Textbook pages 14–15

NAME:

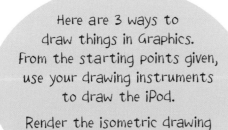

Here are 3 ways to draw things in Graphics. From the starting points given, use your drawing instruments to draw the iPod.

Render the isometric drawing with colouring pencils.

3rd Angle Orthographic Projection (2D)
Looking square on at the front (front elevation), sides (end elevation) and down onto the top (plan).

Oblique (3D)
(Oblique means angle)
Looking square on at the front and down onto the top and one side.

Isometric (3D)
(Isometric means equal angles)
Looking at an angle at two sides and down onto the top.

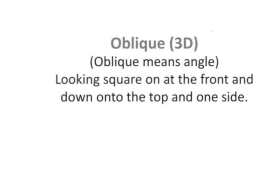

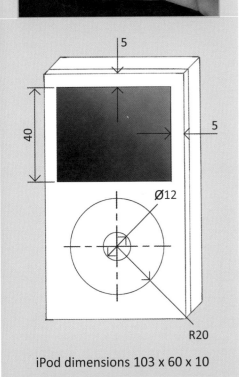

iPod dimensions 103 x 60 x 10

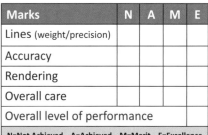

Marks	N	A	M	E
Lines (weight/precision)				
Accuracy				
Rendering				
Overall care				
Overall level of performance				

N=Not Achieved A=Achieved M=Merit E=Excellence

Printing

Textbook page 16

In the boxes given, neatly copy the Graphics poem.

Use UPPER CASE LETTERS and print between the guide lines at the bullet point (•).

Use a very sharp HB pencil.

GRAPHICS IS COOL
GRAPHICS IS NEAT
USING A TEE SQUARE IS REALLY SWEET

THE EQUIPMENT I USE
IS NOT VERY CHEAP
BUT MY LINES AND MY PRINTING
I'LL PRACTISE ON HEAPS

My first attempt at printing was stink!
My second attempt is better I think ...

But wait, here's my last, it's better by far. I'm now an expert at printing ...
HOORAH!

My 1st attempt

My 2nd attempt

My 3rd attempt

Use a pencil sharpener to keep your pencil sharp!

Marks	N	A	M	E
Lines (weight/precision)				
Consistency of lettering				
Correct standards				
Overall care				
Overall level of performance				

N=Not Achieved   A=Achieved   M=Merit   E=Excellence

Beginning Drawing

Textbook pages 18–19

Marks	N	A	M	E
Lines (weight/precision)				
Accuracy				
Title block				
Printing standard				
Overall level of performance				

N=Not Achieved   A=Achieved   M=Merit   E=Excellence

Compass Exercises

Textbook pages 21–22

Marks	N	A	M	E
Compass lines (weight/precision)				
Accuracy				
Title block				
Printing standard				
Rendering				
Overall level of performance				
N=Not Achieved A=Achieved M=Merit E=Excellence				

ISBN 978-0-17-018562-2

Orthographic Projection 1

Textbook pages 23–24

Marks	N	A	M	E
Lines (weight/precision)				
Accuracy				
Title block				
Printing standard				
Overall level of performance				

N=Not Achieved   A=Achieved   M=Merit   E=Excellence

Orthographic Projection 2

Textbook page 27

Marks	N	A	M	E
Lines (weight/precision)				
Accuracy				
Title block				
Printing standard				
Overall level of performance				|//
N=Not Achieved A=Achieved M=Merit E=Excellence				

Solids Rendering

Textbook pages 28–29

The outlines of the four main solids, Cube (incomplete), Cone, Sphere and Cylinder are provided.

1. Place this worksheet onto your drawing board, then with your 30/60° set square and tee square, join the incomplete outside edges of the cube. Where the lines meet, draw the three missing axes (30° and vertical) as shown below. *Make the lines very light*.

2. Working on a flat surface:
   - Complete the tonal scale.
   - Using pencil strokes parallel to the sides of the solids, apply tone to each surface with a 4B pencil. (The direction of light is from right to left.)
   - Smudge with a tissue and make white highlights with your eraser.
   - Render the shadows the same tone as the darkest surface of each solid.

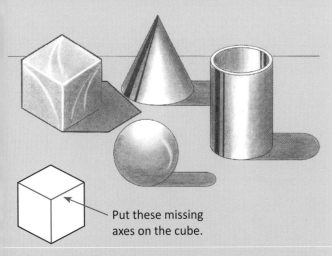

Put these missing axes on the cube.

3. Use the same techniques to render the toy train with colouring pencils.

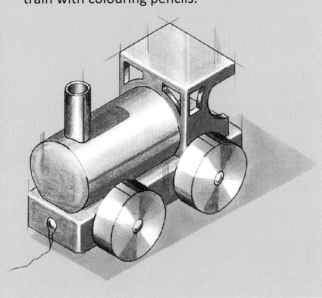

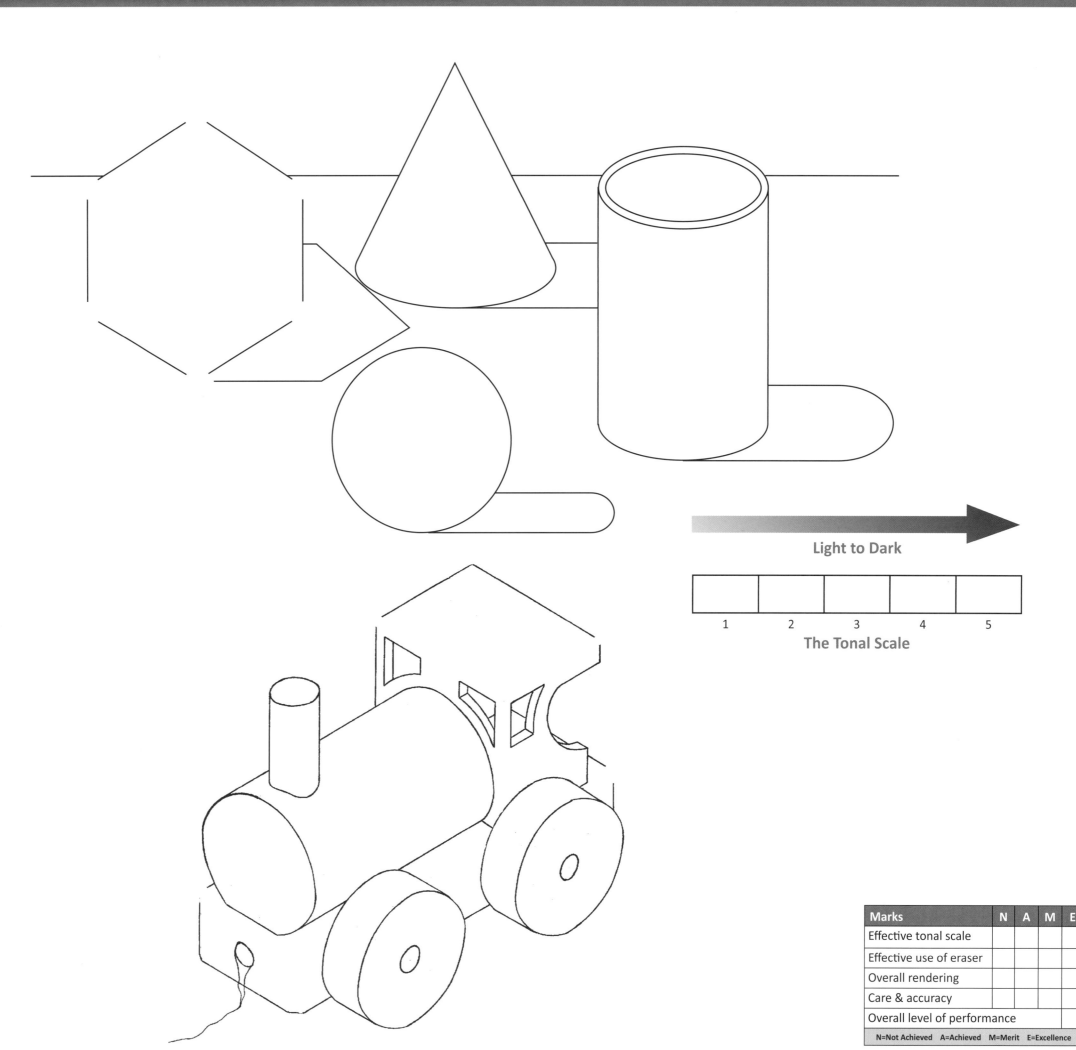

Light to Dark

1	2	3	4	5

The Tonal Scale

Marks	N	A	M	E
Effective tonal scale				
Effective use of eraser				
Overall rendering				
Care & accuracy				
Overall level of performance				

N=Not Achieved A=Achieved M=Merit E=Excellence

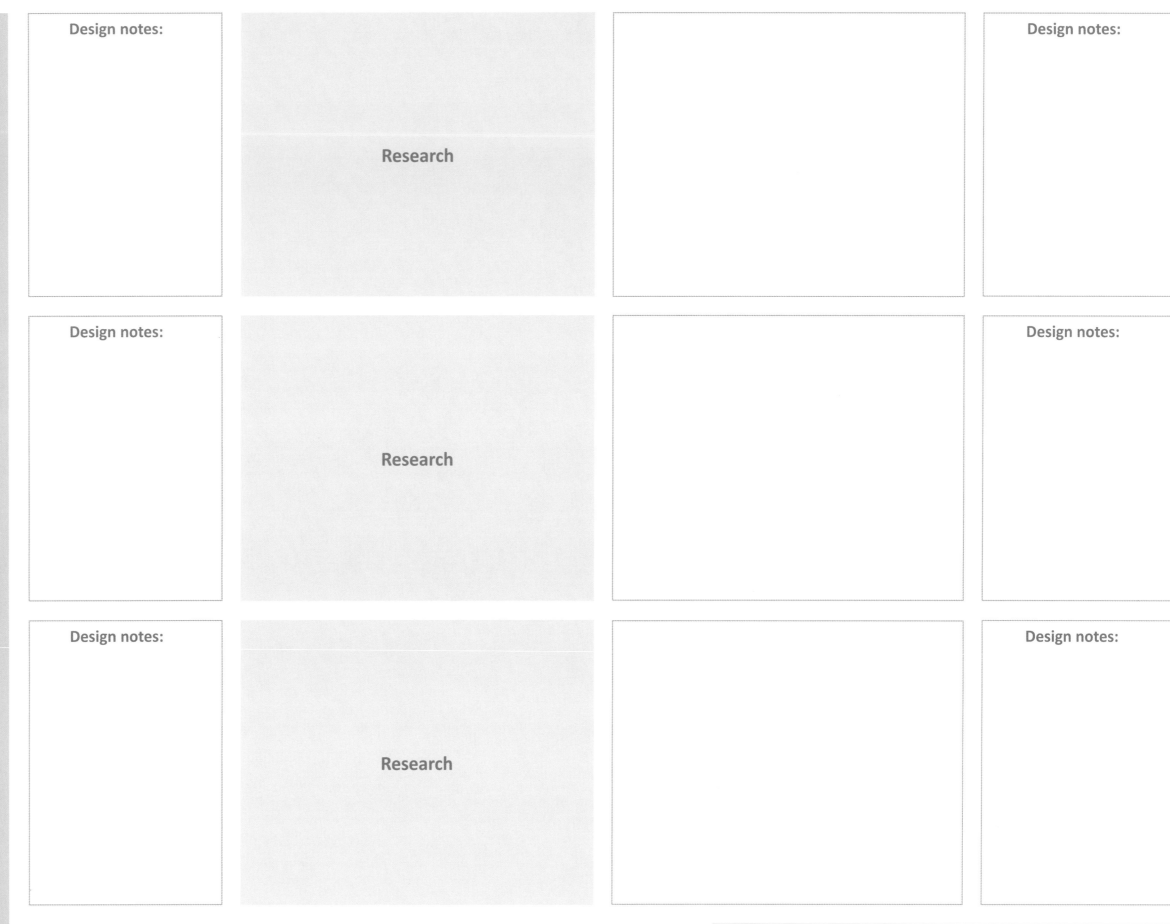

Logo Design

Here is an introduction to the process of design. Design a logo (*Logogram*) to be placed on your design assignment that will identify your work and yourself to others.

1. Collect 3 examples of logos from magazines/newspapers/internet etc. This is called *design research* or *investigation*.

2. Paste your research onto this page in the boxes labelled **Research**.

3. Place brief comments (design notes) beside your research. Say:
   - What you like or dislike about each logo.
   - What you think is significant about any colours or shading in the logo.
   - What is important about the shape or message in the logo.

4. Using ideas from the research, produce one **concept sketch** in each box beside the research (Concept means idea).

5. Place design notes beside each concept. Say:
   - What you like or dislike about each.
   - What is important about the shape or message in each concept.

6. Choose and identify your best logo concept by printing *My Chosen Concept* beneath it.

7. Use these design words (called design language) in your design notes: *style, aesthetics, proportion, colour, harmony*.
   - **Aesthetics** means pleasing to the eye.
   - **Proportion** describes sizes and shapes and their relationships to each other.
   - **Style** is whether or not you think an object or shape is 'in fashion' or ugly or pleasing.
   - **Harmony** describes different parts of a design that blend well together.

Notes
- Use your 4B pencil for sketches.
- Use colour for effect.
- Your logo concepts may have just graphics (images) or be a combination of graphics and text.
- Keep your designs simple.

Marks	N	A	M	E	Marks	N	A	M	E
Research quality					Sketching quality				
Quality of ideas					Design notes & design language				
Overall care					Overall level of performance				

N=Not Achieved   A=Achieved   M=Merit   E=Excellence

3D Freehand Sketching

Textbook pages 30–32

10

1. Use the starting points and a 4B pencil to make freehand 3D sketches of:
   - The Telephone in **isometric**.
   - The Tool Box in **oblique**.
2. Make boxes (called CRATES) first, this is the overall size.
3. Be very accurate and make sure that the **proportions** are correct, then carefully draw in the details of each.
4. Use colouring pencils to render the finished sketches to show tonal values and the materials they are made from.

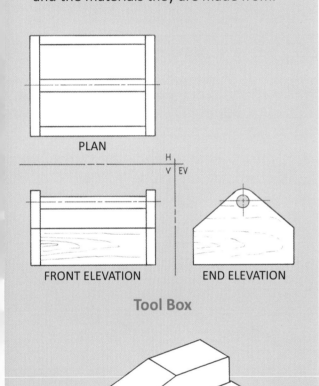

Tool Box

Telephone

Oblique Tool Box

Isometric Telephone

Marks	N	A	M	E
Crating (straight, parallel lines)				
Objects carefully placed in crates				
Effective colour rendering				
Care & accuracy				
Overall level of performance				

N=Not Achieved A=Achieved M=Merit E=Excellence

2D & 3D Freehand Sketching

Textbook page 33

1. Use the starting points and a sharp 4B pencil to make a freehand 2D sketch of the Hall Table and Chair.

2. Each sketch will be an **orthographic projection**. Show a plan, a front elevation and one end elevation. *View the front elevations of each in the direction indicated by the arrows.*

3. Make boxes (CRATES) first. This is the overall sizes. Be very accurate and make sure that the **proportion** is correct, then carefully draw in the details. *Judge all sizes of each object.*

4. Put some rendering on the finished sketches to show materials.

Remember
- Make all lines very light construction to begin with.
- Outline the details of each object with a dark, thin line.
- Leave projection lines light but clear.

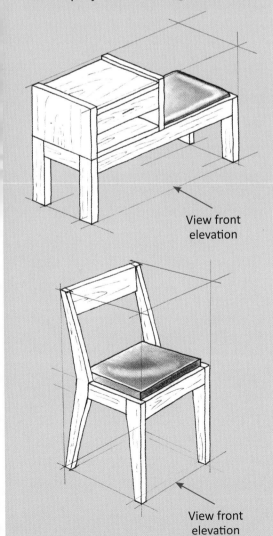

View front elevation

View front elevation

FRONT ELEVATION

FRONT ELEVATION

Hall Table

Chair

Marks	N	A	M	E
Crating (straight, parallel lines)				
Objects carefully placed in crates				
Projection lines				
Care & accuracy				
Overall level of performance				

N=Not Achieved   A=Achieved   M=Merit   E=Excellence

3D Rapid Viz

Textbook page 33

1. The word concept means 'idea'. Concepts are quick, freehand sketches (rapid viz) of an idea, constructed inside crates.

2. Use a 4B pencil to sketch the oblique and isometric objects shown below. Make the sketches large enough to fill the page, side by side. Be sure to make the oblique and isometric axes a low angle, rather than steep.

3. Use quick rendering techniques to show tonal values, shiny and dull surfaces and surface materials such as wood, plastic, metal etc.

4. To heighten the impact of each sketch, show backgrounds and thick and thin lines.
   - Make the lines that show the join between two surfaces, but where only one of those surfaces is seen, a **thick line**.
   - Make the line that shows the join between two surfaces, but where both surfaces can be seen, a **thin line**.

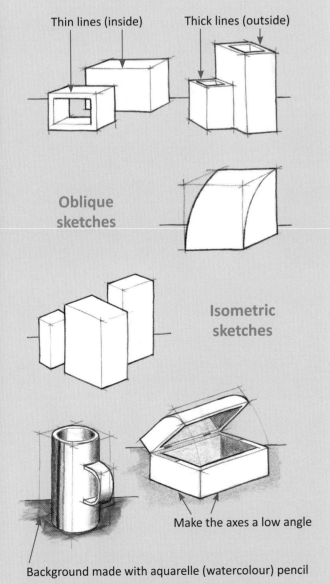

Thin lines (inside) | Thick lines (outside)

Oblique sketches

Isometric sketches

Make the axes a low angle

Background made with aquarelle (watercolour) pencil

Oblique Sketches

Isometric Sketches

Marks	N	A	M	E
Crating (straight, parallel lines)				
Objects carefully placed in crates				
Thick & thin lines				
Rendering				
Overall level of performance				
N=Not Achieved A=Achieved M=Merit E=Excellence				

3D Exploded Sketches
Textbook page 34

Two common wood joints found in furniture and building construction can be seen below.

1. Use the starting points **X** to draw an accurate, well-proportioned freehand exploded isometric sketch of each. Judge the sizes of each part carefully.

2. Use crates and explode in the direction of the arrows (*which is the way the joints would be assembled and/or pulled apart*).

3. Show thick and thin lines and colour render to show wood texture and tonal values.

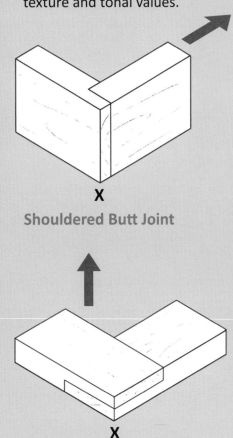

Shouldered Butt Joint

Angled Halving Joint

X
Shouldered Butt Joint

X
Angled Halving Joint

Marks	N	A	M	E
Crating (straight, parallel lines)				
Objects carefully placed in crates				
Care & accuracy				
Overall level of performance				

N=Not Achieved A=Achieved M=Merit E=Excellence

2D Detail Sketches

You can show more clearly what an object looks like inside by imagining it to be cut open.

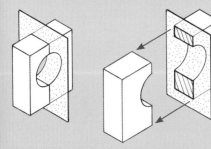

The parts that are cut are shown cross hatched.

The torch from Worksheet 21, and a crank and cam are shown below.

1. Draw a well-proportioned 2D freehand sectioned sketch of each (*show the crank and cam assembled as seen in the front elevation below*).
2. Use crates for good proportion.
3. Show 'hatching' on the surfaces that are imagined to be cut.

Make the sketches larger rather than smaller. For fine detail and cross hatching, use a sharp HB pencil instead of a 4B.

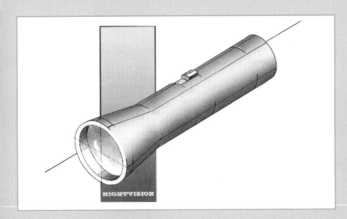

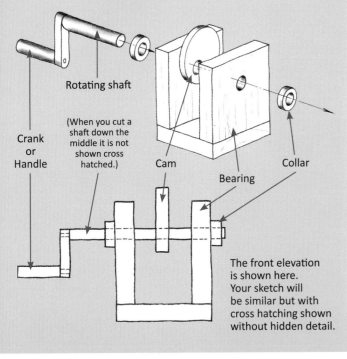

Rotating shaft

(When you cut a shaft down the middle it is not shown cross hatched.)

Crank or Handle | Cam | Bearing | Collar

The front elevation is shown here. Your sketch will be similar but with cross hatching shown without hidden detail.

Marks	N	A	M	E
Crating (straight, parallel lines)				
Objects carefully placed in crates				
Hatching shown				
Overall care & accuracy				
Overall level of performance				
N=Not Achieved A=Achieved M=Merit E=Excellence				

Oblique Drawing

Textbook pages 35–38

Marks	N	A	M	E
Lines (weight/precision)				
Title block				
Printing standard				
Overall accuracy				
Overall level of performance				|//
N=Not Achieved A=Achieved M=Merit E=Excellence				

Isometric Drawing

Textbook pages 39–42

Marks	N	A	M	E
Lines (weight/precision)				
Title block				
Printing standard				
Overall accuracy				
Overall level of performance				

N=Not Achieved   A=Achieved   M=Merit   E=Excellence

Isometric Circles and Curves
Textbook pages 43–44

NAME:

Exercise 1: Isometric Cylinder
Use the **Compass Method** of circle and curve construction to complete the exercises.

1. Using the starting point given, draw the cylinder shown below in isometric when looking down on it. *Draw the bottom of the cylinder first.*

Isometric Cylinder
Cylinder: Ø60
Axis: 80

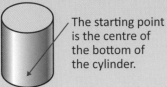
The starting point is the centre of the bottom of the cylinder.

2. Complete the isometric and oblique suitcases to show the missing rounded corners.
   - **Corners R65**
   - **Scale 1:5**

Exercise 2: Suitcases
Hint: Extend the given edges of the suitcases first (the axes), to make the missing rounded corners straight. Then set your compass to the radius of the curve and step it along the axes from the corner.

Rule: A circle will fit inside a square.

Centre lines travelling from the side of the square where the circle touches form a right angle (90°). Where these lines meet is the compass point for drawing the circle (at C below).

Follow this rule for locating the compass point for curves when they are on an angle.

Method
1. Draw the sides of the box by extending the axes to make a square end of straight lines (**A**).
2. Set your compass to the radius of the curve and step this along the straight lines from the corner. You now have the point on the lines where the curve will touch (**B**).
3. At this point make a right angle (90°). Where these 90° lines meet is the compass point (**C**).
4. Put your compass on this point, set it to the length of the line and draw the curve.

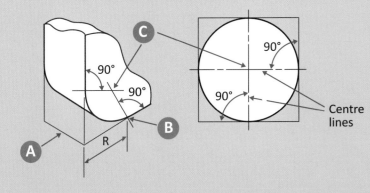

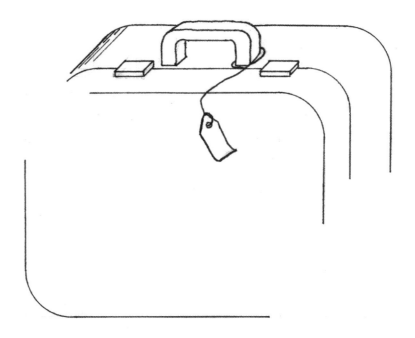
Oblique Suitcase

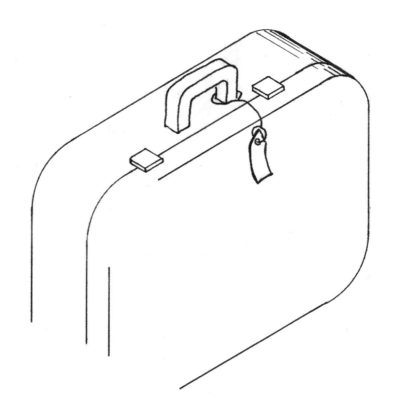
Isometric Suitcase

Isometric Cylinder

Marks	N	A	M	E
Compass curves (weight/precision)				
Lines (weight/precision)				
Centre lines				
Overall accuracy				
Overall level of performance				

N=Not Achieved   A=Achieved   M=Merit   E=Excellence

Isometric Circles – Toy Truck

18

Textbook page 45

NAME:

The front and end elevations of a toy truck are shown.

1. Use the starting point **X** to redraw the toy truck in isometric.
2. Use a compass to transfer all measurements from the given views. (Draw, in light construction, a box about the cab on the front elevation. This will become an *auxiliary view*.)
3. Judge any sizes not given.
4. Use an ellipse template to draw the wheels. The method is described below.
5. Show the truck tray empty. Print the words **TRUCK COMPANY** on the side.
6. Show all constructions lightly but clearly.

Using the ellipse template

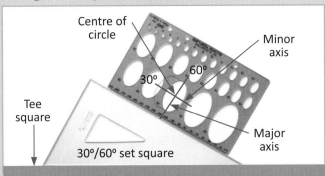

The long axis through the ellipse is called the **MAJOR AXIS (60°)**.

The short axis through the ellipse is called the **MINOR AXIS (30°)**.

Follow the steps below for drawing the wheels of the toy truck:

1. Find the centre of the back of the wheel.
2. From the centre, draw a 30° line (minor axis) and a 60° line (major axis).
3. Draw a half ellipse with the template.
4. Draw a line at 30° from the centre of the back of the wheel and step the wheel thickness along it.
5. Draw a 30° and a 60° line. (Another major and minor axis.)
6. Draw a full ellipse with the template.
7. Join the two ellipses with 30° lines to form the sides of the wheel.
8. Draw CENTRE LINES through the front of the wheels (vertical and 30°).

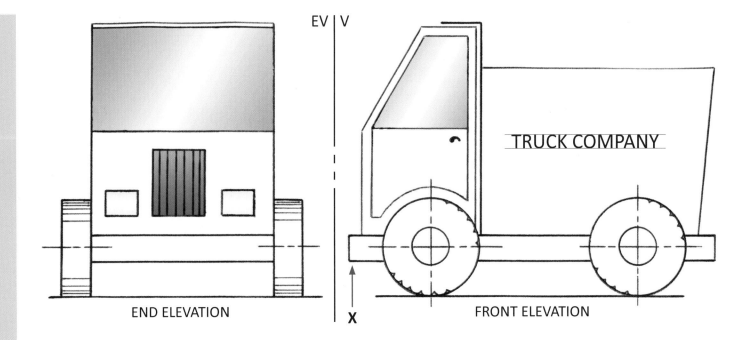

END ELEVATION   EV | V   FRONT ELEVATION

X

Marks	N	A	M	E
Lines (weight/precision)				
Accurate plotting from elevations				
Construction of wheels				
Isometric printing				
Overall accuracy				
Overall level of performance				
N=Not Achieved A=Achieved M=Merit E=Excellence				

One Point Perspective

Textbook pages 46–49

Exercise 1: Camera
A rendered one point perspective drawing of a digital camera is shown below.

1. Between the guide lines given, use a sharp HB pencil to neatly label the horizon line and vanishing point (HORIZON LINE, VP).

2. Use the starting points to draw the one point perspective camera as shown.

3. Render the drawing to make it look as real as possible.

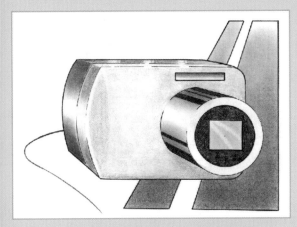

Exercise 2: Fence
The front elevation of a fence is shown below. The centre line and height of the first fence post is also given.

1. Between the guide lines given, use a sharp HB pencil to neatly label the horizon line and vanishing point (HORIZON LINE, VP).

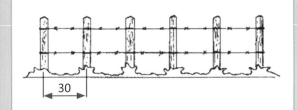

30

2. Beginning at the centre line of the first post, redraw the fence post in one point perspective.

3. Render the drawing to show the fence in a landscape.

Skills
Drawing circles in perspective and finding equal spaces when given the distance between equal spaces.

Digital Camera

Fence

Marks	N	A	M	E
Lines (weight/precision)				
Printing standard				
Accuracy of the drawings				
Rendering				
Overall care				
Overall level of performance				
N=Not Achieved A=Achieved M=Merit E=Excellence				

Two Point Perspective

Textbook pages 50–53

Exercise 1: Kitchen Cupboard
A kitchen cupboard attached to the wall and a sink unit is given.

1. Between the guide lines given, use a sharp HB pencil to neatly label the vanishing points VP1 (*left*) VP2 (*right*).

2. Using the construction method of equal space division, place four equally spaced doors on the front of the cupboard unit.

3. Add any other features to the drawing then render to make it look as real as possible.

Exercise 2: Cottage
A small cottage is shown below. Starting points (dots) that need to be joined with visual rays that travel to the vanishing points, are also provided.

1. Join the starting dots with visual rays that travel to the vanishing points to form the outline of the cottage.

2. Continue the visual rays to locate VP1 on the left side of the drawing.

3. Between guide lines use a sharp HB pencil to neatly label the vanishing points VP1 (*left*) VP2 (*right*).

4. Place windows, door, a deck, pathway and landscaping features onto the drawing. Judge all sizes and render the drawing.

Skills
Finding equal spaces when given the distance to divide, finding centres, simple landscape graphics.

Marks	N	A	M	E
Lines (weight/precision)				
Printing standard				
Accuracy of the drawings				
Rendering				
Overall care				
Overall level of performance				

N=Not Achieved   A=Achieved   M=Merit   E=Excellence

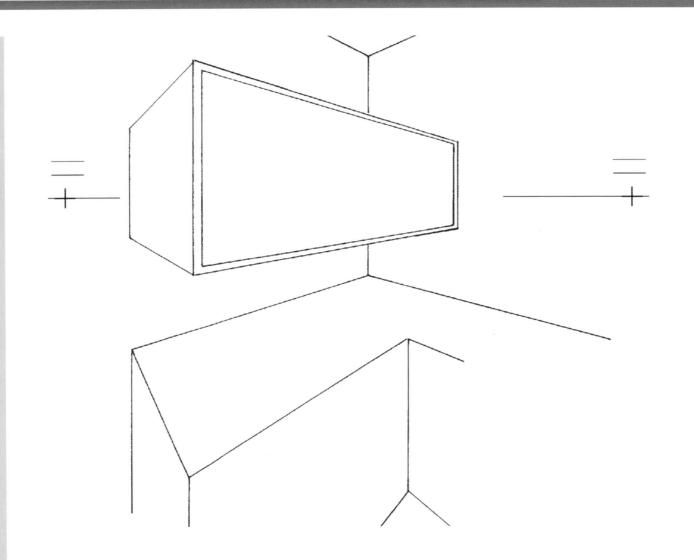

Kitchen Cupboard

Cottage

Isometric Circles — Torch

Use the skills you have learnt earlier and the instructions below to complete the isometric drawing of a torch.

Method

1. Complete the Ø60 outside circular end of the torch. *The method is the same as you used for the cylinder on Worksheet 18, but on its side.*

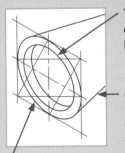

The inside circular end of the torch has been drawn for you.

Continue the outside lines of the sides of the torch to meet the circular end.

The outside circular end of the torch is Ø60. You will need to draw this part.

2. Extend the outside edges of the torch to meet the circular end.

3. Render the torch. Use the airbrush if you have access to one or choose a medium of your own choice. Think carefully about how light falls on the curved surface of a cylinder.

4. Outline the outside edges of the torch.

5. Cut this page down the dashed line then carefully cut out your completed torch. *Take care not to cut into the outlines of the torch.*

6. Design and make a background on which to paste the torch. A simple rectangle is often the best idea for products.

7. Paste the torch onto the background, making sure that the vertical centre line remains vertical.

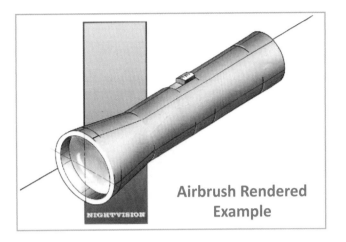

Airbrush Rendered Example

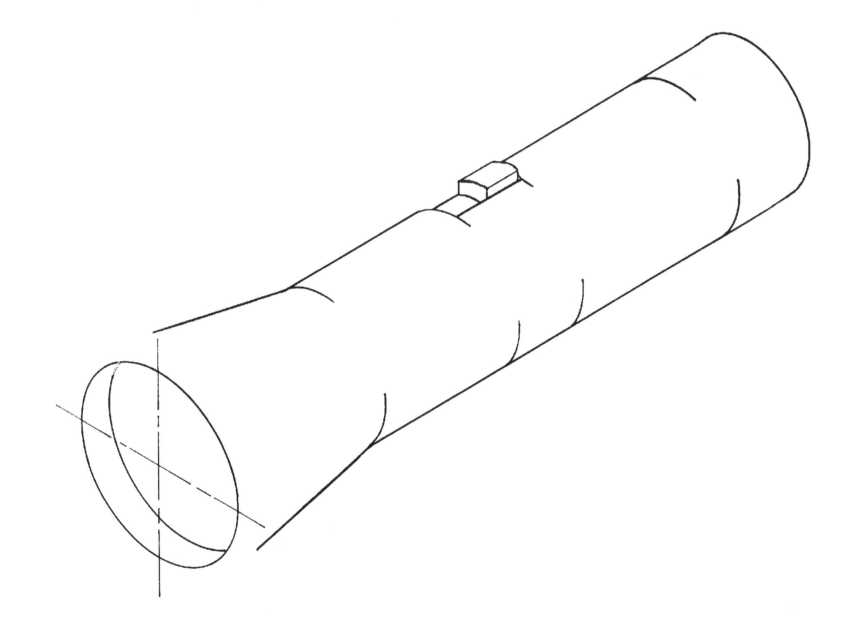

Marks	N	A	M	E
Compass lines (weight/precision)				
Rendering				
Background				
Overall care & accuracy				
Overall level of performance				

N=Not Achieved   A=Achieved   M=Merit   E=Excellence

Geometry – Polygons

Regular Hexagon

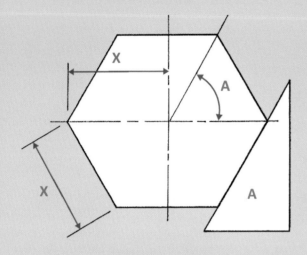

Do you know what the lengths marked **X** have in common?

Do you know how many degrees are in the angles marked **A**? Why?

Constructed inside an Ø80 circle.

Constructed on a given line.

Constructed about a Ø78 circle (78 A/F).

Regular Octagon

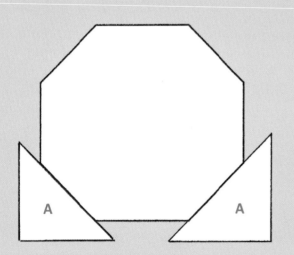

Do you know how many degrees there are in the angle marked **A**? Why?

Constructed inside an 80 mm square.

Constructed inside an Ø80 circle.

Constructed about a Ø78 circle (78 A/F).

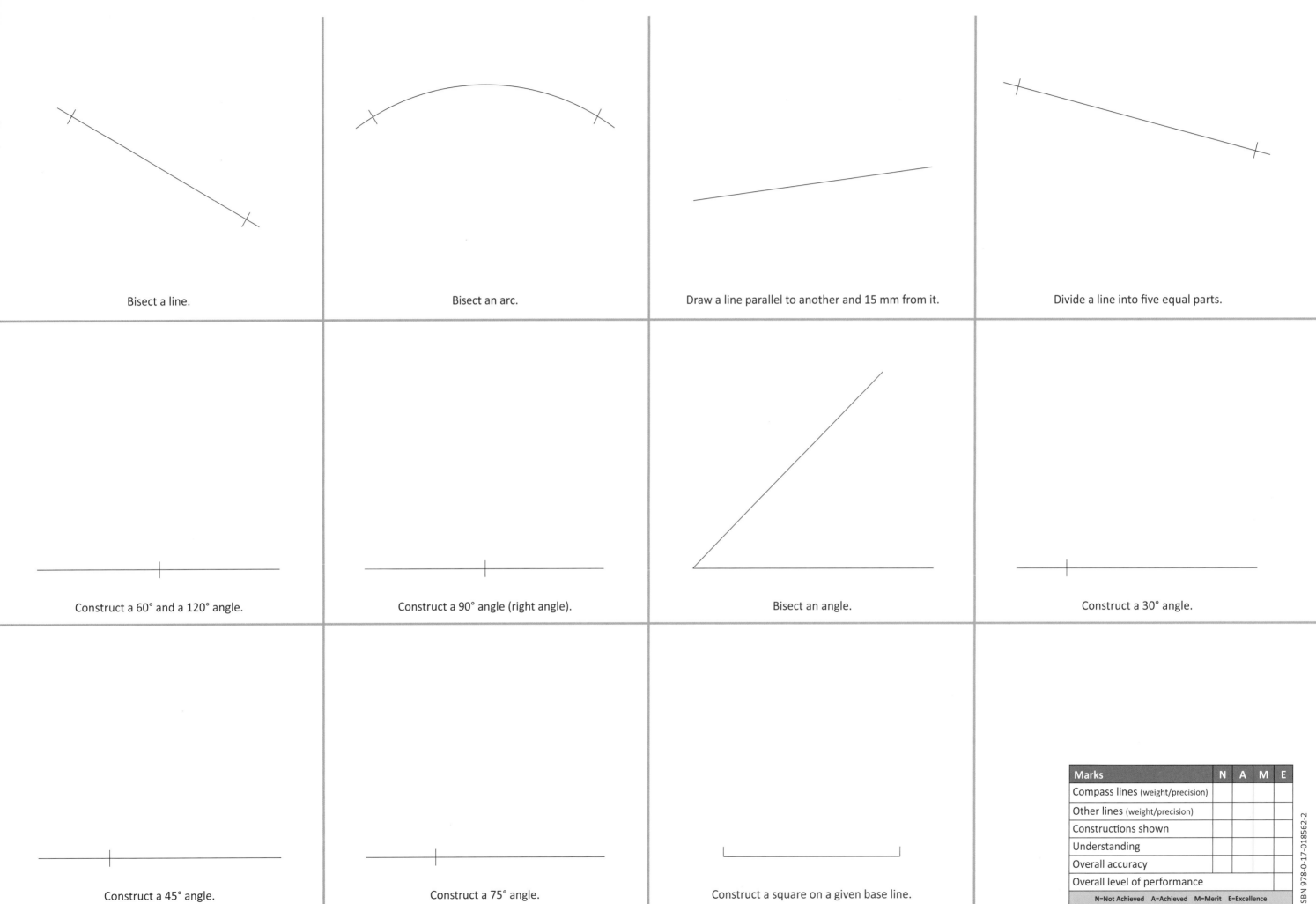

Angles in Isometric and Line Division

Textbook page 57 — 24

The orthographic projections of a salt shaker and dog kennel are given.

1 Use the starting points **X** to redraw each in isometric. Show all constructions clearly. *Draw a crate in light construction around the elevations of the orthographic projections first, then redraw this crate the same size in isometric (see below).*

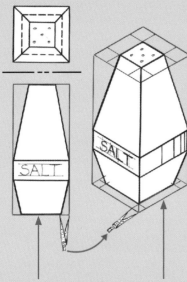

Draw a crate around the elevations then redraw this crate the **same size** in isometric.

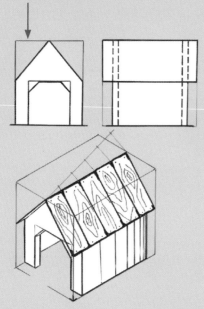

2 Use dividers or a sharp compass to transfer measurements (called plotting) from the axes of the orthographic views crate to the isometric crate.

3 Write the word SALT on the isometric salt shaker and show equal spaces construction for the boards on the roof of the dog kennel.

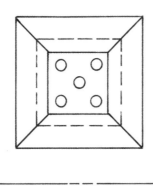

H / V

X

Direction of view

Construct with instruments, five equally spaced boards on the roof of the isometric drawing.

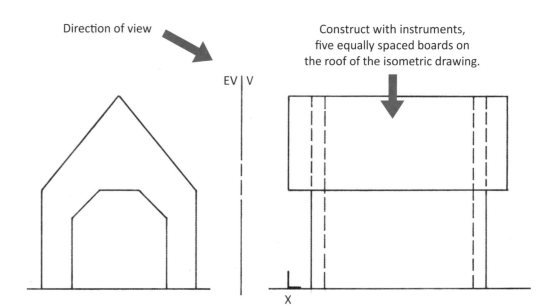

X

Marks	N	A	M	E
Lines (weight/precision)				
Accuracy of drawing				
Overall care				
Overall level of performance				

N=Not Achieved A=Achieved M=Merit E=Excellence

Isometric Chocolate Bar Box

Year 9 GRAPHICS — Textbook pages 58–59 — 25

Marks	N	A	M	E
Lines (weight/precision)				
Auxiliary view drawn (hexagon)				
Accurate plotting from aux. view				
Concepts and rendering				
Title block				
Printing standard				
Overall level of performance				
N=Not Achieved A=Achieved M=Merit E=Excellence				

ISBN 978-0-17-018562-2